Ecosystem Restoration

END PLASTICS
2024

Sophia Reynolds

All rights reserved. No part of this publication may be reproduced, distributed, or transmitted in any form or by any means, including photocopying, recording, or other electronic or mechanical methods, without the prior written permission of the publisher, except in the case of brief quotations embodied in critical reviews and certain other noncommercial uses permitted by copyright law.

TABLE

OF

CONTENTS

Setting the stage 1
The global plastic crisis 1
Importance of ecosystem restoration 4

Understanding Plastic Pollution 6
Origins of plastic pollution 6
Impact on ecosystems 9
Human health implications 11

The Role of Plastic in Modern Society 13
Historical context of plastic production 13
Current usage and dependence on plastic 16
Economic implications of the plastic industry 19

Principles and Practices 22
Significance of ecosystem restoration 22
Successful restoration efforts 25
Successful restoration projects 28

Challenges to Ecosystem Restoration 31
Political and economic obstacles 31
Technological limitations 34
Social and cultural barriers 38

Innovative Solutions to End Plastic Pollution 42
Biodegradable alternatives to traditional plastics 42
Technological advancements in waste management 47
Policy recommendations for reducing plastic usage 51

Collaborative Efforts and Community Engagement ... 54
Importance of interdisciplinary collaboration 54

Engaging communities in restoration efforts 57
Grassroots initiatives and citizen science 60
The Future of Plastic-Free Ecosystems 64
Vision for a plastic-free future 64
Potential obstacles and solutions 68
Call to action for individuals and organizations 72
Urgency in ending plastic pollution 76

Welcome

In the modern era, humanity stands at a crossroads, facing one of its greatest challenges: the pervasive and escalating crisis of plastic pollution. As plastic production and consumption continue to surge, our planet's ecosystems are buckling under the weight of plastic waste, threatening biodiversity, contaminating waterways, and jeopardizing the health and well-being of present and future generations.

"End Plastics: Ecosystem Restoration" is a call to action, a manifesto for change, and a roadmap for building a sustainable future free from the scourge of plastic pollution. In these pages, we embark on a journey to explore the interconnected issues of plastic pollution and ecosystem degradation, and to chart a course towards restoration, renewal, and resilience.

From the origins of plastic production to the far-reaching impacts on ecosystems and human health, we delve into the complexities of the global plastic crisis and confront the urgent need for transformative solutions. Drawing on scientific research, policy analysis, and real-world examples, we examine the multifaceted dimensions of plastic pollution and its profound implications for the environment, society, and economy.

But amidst the sobering realities of plastic pollution, there is also hope – hope born from the resilience of nature, the ingenuity of human innovation, and the power of collective action. As we confront the challenges of our time, we are reminded of our shared responsibility as stewards of the planet, entrusted with the task of safeguarding its precious ecosystems and biodiversity for future generations.

"End Plastics: Ecosystem Restoration" is more than a book – it is a manifesto for change, a blueprint for action, and a rallying cry for a global movement dedicated to ending plastic pollution and restoring balance to our planet. Together, let us rise to the challenge, embrace the opportunity, and embark on a journey towards a sustainable future where ecosystems thrive, plastic pollution is a distant memory, and humanity lives in harmony with nature.

Chapter One

Setting the stage

The global plastic crisis

The global plastic crisis represents one of the most pressing environmental challenges of our time, stemming from the widespread production, consumption, and disposal of plastic materials. Here's an in-depth look at the various aspects of this crisis:

1. **Production and Consumption:** The production of plastic has skyrocketed since its inception in the mid-20th century. Currently, over 350 million metric tons of plastic are produced annually worldwide. This surge is driven by its versatility, affordability, and durability, making it a preferred material for various industries such as packaging, construction, healthcare, and electronics. However, this

widespread use also contributes significantly to the accumulation of plastic waste.
2. **Disposal and Waste Management**: A major challenge of the plastic crisis lies in its disposal. A considerable portion of plastic products is designed for single-use purposes, leading to a rapid accumulation of waste. Improper disposal practices, such as littering and inadequate waste management infrastructure, exacerbate the problem. Much of this waste ends up in landfills, where it can take hundreds to thousands of years to decompose. Additionally, a significant amount of plastic waste finds its way into natural environments, particularly oceans and waterways.
3. **Impact on Ecosystems**: Plastic pollution poses a severe threat to ecosystems worldwide. Marine environments, in particular, bear the brunt of plastic pollution, with an estimated 8 million metric tons of plastic entering the oceans annually. This pollution harms marine life through entanglement, ingestion, and habitat destruction. Species ranging from seabirds and turtles to fish and whales suffer from the adverse effects of plastic pollution, leading to population declines and ecosystem disruption. Furthermore, plastic pollution extends beyond marine environments, affecting terrestrial ecosystems and freshwater habitats as well.
4. **Human Health Implications**: The plastic crisis also has significant implications for human health. Plastics contain various chemicals, such as phthalates, bisphenol A (BPA), and polybrominated diphenyl ethers (PBDEs), which can leach into the environment and accumulate in the food chain. Studies have linked exposure to these chemicals with adverse health effects, including reproductive disorders, developmental abnormalities, and certain cancers. Additionally, microplastics, tiny plastic particles less than 5mm in size, have been found in drinking water, seafood, and even the air we breathe, raising concerns about their potential impact on human health.
5. **Economic and Social Costs**: The plastic crisis imposes significant economic and social costs on communities and nations worldwide. Cleanup and mitigation efforts require substantial financial resources, placing a burden on taxpayers and local governments. Plastic pollution also affects industries such as tourism and fisheries, which rely on healthy ecosystems for their livelihoods. Furthermore,

marginalized communities often bear the brunt of plastic pollution, particularly in developing countries with inadequate waste management infrastructure.
6. **Climate Change and Environmental Justice**: The production and disposal of plastic also contribute to climate change. The extraction of fossil fuels for plastic production, along with the energy-intensive manufacturing processes, generates greenhouse gas emissions. Additionally, plastic pollution exacerbates climate change by disrupting marine ecosystems, which play a crucial role in carbon sequestration and regulating the Earth's climate. Addressing the plastic crisis requires a holistic approach that considers environmental justice, ensuring that solutions prioritize the needs of frontline communities disproportionately affected by plastic pollution.

Importance of ecosystem restoration

Ecosystem restoration plays a crucial role in combating plastic pollution by addressing both the root causes and the consequences of plastic waste accumulation. Here's a detailed look at the importance of ecosystem restoration in this context:

1. **Mitigating Plastic Pollution at its Source**: Ecosystem restoration helps to mitigate plastic pollution at its source by addressing the underlying drivers of plastic production and consumption. By restoring natural habitats such as forests, wetlands, and mangroves, we can create healthier ecosystems that provide essential ecosystem services, including water filtration, flood control, and carbon sequestration. These restored ecosystems can serve as natural buffers, reducing the flow of plastic waste into waterways and oceans. Additionally, promoting biodiversity and ecological resilience can help to regulate ecosystems and reduce the impacts of plastic pollution on wildlife.
2. **Filtering and Trapping Plastic Waste**: Healthy ecosystems act as natural filters, trapping and retaining plastic waste before it reaches sensitive habitats such as rivers, lakes, and oceans. Mangrove forests, for example, are highly effective at trapping plastic debris and preventing it from entering marine environments. Similarly, coastal dunes and salt marshes can serve as barriers, intercepting plastic waste carried by coastal currents and storm surges. By restoring and protecting these ecosystems, we can enhance their capacity to intercept and sequester plastic pollution, preventing its accumulation in vulnerable habitats.
3. **Promoting Biodegradation and Decomposition**: Ecosystem restoration can facilitate the biodegradation and decomposition of plastic waste through natural processes. Microorganisms present in healthy soils and aquatic environments play a crucial role in breaking down organic matter, including certain types of plastic. By restoring degraded ecosystems and enhancing soil health, we can create conditions that support microbial activity and accelerate the decomposition of plastic materials. Additionally, restoring riparian buffers and vegetative cover along waterways can help to filter out plastic particles and

promote their breakdown through physical and biological processes.
4. **Supporting Circular Economy Solutions**: Ecosystem restoration aligns with principles of the circular economy by promoting sustainable resource management and waste reduction. Restored ecosystems can provide valuable resources such as timber, biomass, and natural fibers, reducing the demand for virgin plastic materials. Furthermore, by integrating ecosystem restoration with sustainable land-use practices such as agroforestry and regenerative agriculture, we can create closed-loop systems that minimize waste generation and maximize resource efficiency. This holistic approach to ecosystem management fosters resilience and sustainability while mitigating the impacts of plastic pollution on ecosystems and communities.
5. **Fostering Community Engagement and Stewardship**: Ecosystem restoration projects provide opportunities for community engagement and stewardship, empowering local residents to take an active role in environmental conservation and plastic pollution mitigation. By involving communities in restoration activities such as tree planting, shoreline cleanups, and habitat restoration, we can raise awareness about the impacts of plastic pollution and foster a sense of ownership and responsibility for ecosystem health. Community-based approaches to ecosystem restoration also help to build social cohesion, resilience, and adaptive capacity, strengthening local communities' ability to address environmental challenges collectively.

Chapter Two

Understanding Plastic Pollution

Origins of plastic pollution

The origins of plastic pollution trace back to the mid-20th century when the mass production of plastics began. Here's a detailed look at the origins of plastic pollution:

1. **Development of Synthetic Polymers**: The development of synthetic polymers in the early 20th century revolutionized manufacturing by providing a versatile and inexpensive material that could be molded into virtually any shape. Bakelite, the first synthetic plastic, was invented in 1907, followed by the creation of other polymers such as polyethylene, polypropylene, and polystyrene. These synthetic polymers offered advantages over traditional materials like wood, metal, and glass, leading to their widespread adoption across various industries.
2. **Rise of Single-Use Plastics**: In the decades following World War II, there was a significant increase in the production and consumption of plastics, driven by economic growth, technological advancements, and changing consumer preferences. One of the key innovations during this period was the development of single-use plastics, including disposable packaging, containers, and utensils. These convenient and inexpensive products revolutionized the way goods were packaged, stored, and transported, but they also contributed to a surge in plastic waste generation.
3. **Expansion of the Plastics Industry**: The plastics industry expanded rapidly throughout the latter half of the 20th century, fueled by advancements in polymer chemistry, manufacturing processes, and marketing strategies. Plastics became ubiquitous in everyday life, with applications

ranging from food packaging and consumer goods to automotive parts and construction materials. This exponential growth in plastic production led to a corresponding increase in plastic waste generation, as many plastic products were designed for single-use purposes and lacked viable recycling options.

4. **Inadequate Waste Management Infrastructure**: The proliferation of plastic products outpaced the development of waste management infrastructure capable of handling the growing volume of plastic waste. In many regions, particularly in developing countries, inadequate waste collection, recycling, and disposal systems have led to widespread littering, illegal dumping, and open burning of plastic waste. As a result, significant quantities of plastic waste end up in natural environments, including oceans, rivers, and terrestrial ecosystems, where they persist for decades to centuries, causing environmental harm.

5. **Plastic in the Marine Environment**: Plastic pollution in the marine environment is of particular concern due to its widespread distribution and persistent nature. An estimated 8 million metric tons of plastic enter the oceans each year, originating from coastal communities, rivers, shipping activities, and offshore platforms. Once in the marine environment, plastic debris can travel vast distances through ocean currents, accumulating in remote areas such as ocean gyres and polar regions. Marine animals, including seabirds, turtles, marine mammals, and fish, are susceptible to ingesting or becoming entangled in plastic debris, leading to injury, suffocation, and death.

6. **Globalization and Transboundary Impacts**: Plastic pollution is a global issue with transboundary impacts, as plastic waste can travel across international borders through marine currents, air currents, and global trade networks. The interconnectedness of the global economy means that plastic pollution generated in one region can affect ecosystems and communities thousands of miles away. Addressing the origins of plastic pollution requires international cooperation, coordination, and collective action to reduce plastic production, promote sustainable consumption, and improve waste management practices on a global scale.

The origins of plastic pollution can be traced to the rapid expansion of the plastics industry, the proliferation of single-use plastics, and inadequate waste management infrastructure. Addressing this complex issue requires a multifaceted approach that involves reducing plastic production, promoting sustainable alternatives, improving waste management systems, and raising awareness about the environmental impacts of plastic pollution.

Impact on ecosystems

The impact of plastic pollution on ecosystems is profound and far-reaching, affecting a wide range of habitats and species across the planet. Here's a detailed exploration of the impact of plastic pollution on ecosystems:

1. **Marine Ecosystems**:
    - **Entanglement**: Marine animals such as seabirds, turtles, seals, and whales often become entangled in plastic debris, leading to injuries, amputations, and even death. Fishing gear, including nets, lines, and traps, poses a significant entanglement risk to marine life.
    - **Ingestion**: Marine animals frequently mistake plastic debris for food, leading to ingestion and subsequent health problems. Plastic particles can block digestive tracts, causing starvation, internal injuries, and malnutrition. Seabirds, sea turtles, fish, and marine mammals are particularly vulnerable to plastic ingestion.
    - **Habitat Degradation**: Plastic pollution alters marine habitats by smothering benthic organisms, covering coral reefs, and suffocating seagrass beds. Microplastics, small plastic particles less than 5mm in size, can accumulate in sediment layers, disrupting nutrient cycling and sediment dynamics. Plastic debris also provides a substrate for invasive species, altering ecosystem structure and function.
    - **Chemical Contamination**: Plastics contain various chemical additives and pollutants, including phthalates, bisphenol A (BPA), and polychlorinated biphenyls (PCBs), which can leach into the marine environment. These chemicals can accumulate in the tissues of marine organisms, biomagnifying through the food chain and posing risks to ecosystem health and human health through seafood consumption.
2. **Freshwater Ecosystems**:
    - **Riverine Pollution**: Rivers and streams serve as conduits for transporting plastic waste from inland areas to coastal environments. Plastic debris can accumulate along riverbanks, clog waterways, and

impede water flow, leading to flooding and habitat degradation. Aquatic organisms, including fish, amphibians, and invertebrates, are susceptible to the impacts of plastic pollution in freshwater ecosystems.
- **Microplastic Contamination**: Microplastics are ubiquitous in freshwater environments, posing risks to aquatic organisms and ecosystem health. Microplastic particles can be ingested by filter-feeders, benthic organisms, and planktonic species, potentially disrupting feeding behaviors, reproductive cycles, and immune responses. Additionally, microplastics can adsorb and transport pollutants, such as heavy metals and persistent organic pollutants, exacerbating contamination levels in freshwater ecosystems.

3. **Terrestrial Ecosystems**:
   - **Soil Contamination**: Plastic pollution affects terrestrial ecosystems through the accumulation of plastic debris in soils and sediments. Improper disposal of plastic waste, including plastic bags, packaging materials, and agricultural plastics, can lead to soil contamination and degradation. Plastic debris can also interfere with soil moisture levels, nutrient cycling, and microbial activity, impacting plant growth and ecosystem functioning.
   - **Wildlife Impacts**: Terrestrial wildlife, including mammals, birds, reptiles, and insects, can be adversely affected by plastic pollution. Animals may become entangled in plastic debris, ingest plastic particles, or use plastic materials in nesting and foraging behaviors. Plastic ingestion can lead to internal injuries, digestive blockages, and nutrient deficiencies, affecting the health and survival of wildlife populations.
   - **Habitat Fragmentation**: Plastic pollution contributes to habitat fragmentation and degradation in terrestrial ecosystems. Plastic debris can obstruct animal migration routes, disrupt wildlife corridors, and fragment natural landscapes. Fragmented habitats are more susceptible to invasive species, habitat loss, and biodiversity decline, further exacerbating the ecological impacts of plastic pollution.

Human health implications

The human health implications of plastic pollution are a growing concern, as evidence continues to mount regarding the potential risks posed by exposure to plastic materials and associated chemicals. Here's a detailed exploration of the human health implications of plastic pollution:

1. **Chemical Exposure**:
   - **Endocrine Disruption**: Many plastics contain chemicals known as endocrine-disrupting compounds, such as phthalates, bisphenol A (BPA), and flame retardants. These chemicals can interfere with the body's hormone systems, potentially leading to developmental abnormalities, reproductive disorders, and metabolic disturbances. Pregnant women, infants, and children are particularly vulnerable to the effects of endocrine-disrupting chemicals, as exposure during critical periods of development can have long-lasting impacts on health and well-being.
   - **Carcinogenicity**: Some plastic additives and contaminants have been classified as carcinogens or suspected carcinogens, posing risks of cancer development upon long-term exposure. For example, polycyclic aromatic hydrocarbons (PAHs) and vinyl chloride monomers (VCMs) are chemicals found in certain plastics that have been linked to increased cancer risk in humans. Inhalation or ingestion of airborne particles or leachates containing these carcinogens can pose significant health risks, particularly for individuals working in plastic manufacturing facilities or living near waste disposal sites.
2. **Microplastic Ingestion**:
   - **Contaminated Food and Water**: Microplastics, tiny plastic particles less than 5mm in size, are pervasive

in the environment and have been detected in various food and water sources, including seafood, drinking water, and air. Human exposure to microplastics occurs primarily through ingestion, inhalation, and dermal contact. Microplastics can adsorb and concentrate chemical pollutants, such as heavy metals and organic contaminants, which may pose additional health risks upon ingestion.
    - **Gastrointestinal Health**: Microplastic ingestion has the potential to cause gastrointestinal irritation, inflammation, and damage, as plastic particles can accumulate in the digestive tract and interfere with nutrient absorption and gut microbiota balance. Studies have shown that microplastics can translocate across intestinal barriers and enter systemic circulation, raising concerns about their potential systemic effects on human health, including immune dysregulation and chronic inflammation.
3. **Respiratory Health:**
    - **Airborne Particulate Matter**: Plastic pollution contributes to airborne particulate matter through processes such as plastic degradation, incineration, and mechanical abrasion. Inhalation of airborne plastic particles can pose respiratory health risks, particularly for individuals with pre-existing respiratory conditions such as asthma, chronic obstructive pulmonary disease (COPD), and bronchitis. Plastic particles may deposit in the respiratory tract, leading to airway inflammation, oxidative stress, and respiratory symptoms.
4. **Direct Contact and Injury:**
    - **Dermal Irritation**: Direct contact with certain types of plastics or plastic additives may cause dermal irritation, allergic reactions, and skin sensitization in susceptible individuals. Plastic materials containing abrasive additives or chemical accelerators used in manufacturing processes can exacerbate skin conditions such as dermatitis and eczema. Occupational exposure to plastics in industries such as plastics manufacturing, recycling, and waste management may increase the risk of skin disorders and injuries.

Chapter Three

The Role of Plastic in Modern Society

Historical context of plastic production

The historical context of plastic production is a fascinating journey that spans over a century, marked by innovation, technological advancements, and societal transformation. Here's a detailed exploration of the historical context of plastic production:

1. **Early Beginnings (19th Century):**
    - The development of plastics can be traced back to the mid-19th century, with the invention of celluloid, the first synthetic polymer, by John Wesley Hyatt in 1869. Celluloid, derived from cellulose nitrate, was initially used as a substitute for ivory in billiard balls, combs, and photographic film.
    - Bakelite, the first truly synthetic plastic, was invented by Leo Baekeland in 1907. Bakelite, made from phenol and formaldehyde, revolutionized manufacturing with its heat-resistant, durable, and moldable properties. It found applications in electrical insulators, automotive parts, and consumer goods, paving the way for the modern plastics industry.
2. **Expansion and Diversification (20th Century):**
    - The early 20th century witnessed the rapid expansion of the plastics industry, fueled by advancements in polymer chemistry, manufacturing processes, and commercial applications. New types of plastics, such as polyethylene, polyvinyl chloride (PVC), and polystyrene, were developed, each with unique properties and applications.

- World War II played a significant role in accelerating the growth of the plastics industry, as plastics were utilized in military applications such as aircraft components, radar insulation, and protective gear. The war effort spurred innovation in plastics manufacturing techniques, leading to the mass production of lightweight, durable, and versatile materials.
- Post-war economic prosperity and the rise of consumer culture fueled demand for plastic products in civilian markets. Plastics became ubiquitous in everyday life, with applications ranging from packaging and consumer goods to construction materials and medical devices. The 1950s and 1960s are often referred to as the "Golden Age of Plastics," characterized by rapid growth, innovation, and optimism about the potential of synthetic materials to improve quality of life.

3. **Environmental Concerns and Regulation (Late 20th Century):**
   - By the late 20th century, concerns began to emerge about the environmental impacts of plastic production and disposal. Plastic pollution became increasingly visible in the form of littered landscapes, choked waterways, and degraded ecosystems. The durability and persistence of plastics posed challenges for waste management and environmental stewardship.
   - Regulatory efforts to address plastic pollution and mitigate environmental risks were initiated at the national and international levels. Legislation such as the Clean Air Act, the Clean Water Act, and the Resource Conservation and Recovery Act (RCRA) in the United States aimed to regulate emissions, discharges, and disposal of plastic waste.
   - The environmental movement of the 1960s and 1970s brought attention to the need for conservation, pollution prevention, and sustainable resource management. Public awareness campaigns, grassroots activism, and scientific research raised awareness about the ecological impacts of plastic pollution and spurred calls for action to address the growing environmental crisis.

4. **Contemporary Challenges and Innovations (21st Century):**
    - The 21st century presents both challenges and opportunities for the plastics industry, as concerns about plastic pollution, resource depletion, and climate change intensify. The concept of a circular economy, which aims to minimize waste and maximize resource efficiency, has gained traction as a framework for addressing the environmental impacts of plastic production and consumption.
    - Innovations in bioplastics, recycling technologies, and sustainable materials offer promising solutions to reduce the environmental footprint of plastic production and disposal. Biodegradable and compostable plastics derived from renewable sources have emerged as alternatives to traditional petroleum-based plastics, offering potential benefits for waste management and environmental conservation.
    - However, transitioning to a more sustainable plastics economy requires collaboration among stakeholders, including governments, industry leaders, researchers, and consumers. Strategies such as extended producer responsibility (EPR), product stewardship, and eco-design principles can help incentivize producers to adopt more sustainable practices and promote a circular approach to plastic production and consumption.

Current usage and dependence on plastic

The current usage and dependence on plastic are pervasive and multifaceted, influencing virtually every aspect of modern life. Here's a detailed exploration of the current state of plastic usage and dependence:

1. **Packaging and Consumer Goods**:
   - Plastic packaging is one of the largest and most visible applications of plastic, encompassing a wide range of products such as food packaging, beverage containers, household products, and personal care items. Plastics offer versatility, durability, and cost-effectiveness, making them a preferred material for packaging and protecting goods during storage, transportation, and distribution.
   - Single-use plastics, including plastic bags, bottles, straws, and utensils, are widely used in the food and beverage industry, retail sector, and hospitality industry. Despite growing awareness of the environmental impacts of single-use plastics, they remain ubiquitous due to their convenience, affordability, and widespread availability.
2. **Construction and Building Materials**:
   - Plastics are extensively used in the construction industry for a variety of applications, including insulation, piping, roofing, flooring, and windows. PVC, polyethylene, and polystyrene are common types of plastics used in construction materials due to their durability, thermal insulation properties, and resistance to moisture and corrosion.
   - Plastic composites, such as wood-plastic composites (WPCs) and fiber-reinforced plastics (FRPs), are increasingly being used as alternatives to traditional building materials like wood, metal, and concrete. These materials offer advantages such as lower maintenance requirements, longer lifespan, and reduced environmental impact.
3. **Automotive and Transportation**:
   - Plastics play a critical role in the automotive industry, where they are used in vehicle components such as bumpers, dashboards, upholstery, and interior trim.

- Lightweight plastics help to improve fuel efficiency and reduce vehicle emissions, contributing to sustainability and environmental performance.
  - Electric vehicles (EVs) and hybrid vehicles rely on lightweight plastics and composites to optimize energy efficiency and extend driving range. Plastics are also used in battery casings, charging connectors, and thermal management systems for EVs.
4. **Healthcare and Medical Devices**:
   - Plastics are indispensable in the healthcare sector for the production of medical devices, equipment, and packaging materials. Medical-grade plastics such as polypropylene, polyethylene, and polyvinyl chloride (PVC) are used in syringes, IV bags, surgical instruments, and diagnostic equipment.
   - Plastics offer advantages such as biocompatibility, sterility, and ease of sterilization, making them essential for medical applications. Single-use disposable medical devices made from plastics help to reduce the risk of infection and cross-contamination in healthcare settings.
5. **Electronics and Technology**:
   - Plastics are widely used in the electronics and technology industries for the production of electronic components, casings, housings, and displays. Polycarbonate, acrylic, and ABS are common types of plastics used in consumer electronics such as smartphones, laptops, tablets, and televisions.
   - Plastics offer electrical insulation properties, impact resistance, and design flexibility, making them ideal for protecting electronic devices from mechanical damage, moisture, and environmental hazards. Miniaturization and lightweighting trends in electronics drive demand for high-performance plastics with advanced properties.
6. **Agriculture and Packaging**:
   - Plastics play a critical role in modern agriculture for applications such as greenhouse film, mulch film, irrigation systems, and crop protection products. Plastic films help to improve crop yields, conserve water, and reduce pesticide use by providing a controlled environment for plant growth.

- Agricultural plastics such as silage bags, bale wrap, and grain bags are essential for storage, transportation, and preservation of crops and livestock feed. These durable and weather-resistant materials help farmers to optimize harvest efficiency and reduce post-harvest losses.

7. **Textiles and Apparel**:
    - Plastics are increasingly being used in the textile and apparel industry for the production of synthetic fibers, fabrics, and clothing. Polyester, nylon, and acrylic are common types of synthetic fibers derived from petroleum-based plastics.
    - Synthetic fibers offer advantages such as durability, moisture-wicking properties, and colorfastness, making them popular for activewear, sportswear, outdoor gear, and performance apparel. However, concerns have been raised about the environmental impact of microfiber pollution from synthetic textiles.

8. **Food and Beverage Industry**:
    - Plastics are extensively used in the food and beverage industry for packaging, storage, and transportation of food products. Plastic containers, trays, wraps, and films help to preserve freshness, extend shelf life, and prevent contamination of perishable goods.
    - The convenience and cost-effectiveness of plastic packaging have made it the preferred choice for food manufacturers, retailers, and consumers. However, concerns about food safety, chemical migration, and environmental pollution have prompted efforts to reduce plastic usage and promote sustainable alternatives.

Economic implications of the plastic industry

The economic implications of the plastic industry are significant and multifaceted, spanning production, manufacturing, consumption, trade, and waste management. Here's a detailed exploration of the economic aspects of the plastic industry:

1. **Employment and Economic Growth**:
    - The plastic industry is a major contributor to global employment and economic growth, supporting millions of jobs across various sectors, including manufacturing, packaging, transportation, and retail. The industry encompasses a diverse range of businesses, from large multinational corporations to small and medium-sized enterprises (SMEs), providing employment opportunities for workers with diverse skill sets and educational backgrounds.
    - Plastic production and manufacturing contribute to economic development by generating revenue, stimulating investment, and driving innovation. The industry fosters technological advancements, process improvements, and product innovations that enhance productivity, efficiency, and competitiveness.
2. **Value Chain Integration**:
    - The plastic industry is deeply integrated into the global economy, with extensive linkages across multiple sectors and value chains. Plastics serve as essential inputs for numerous industries, including automotive, construction, electronics, healthcare, agriculture, and consumer goods. The versatility and adaptability of plastics make them indispensable for a wide range of applications, driving demand and market expansion.
    - Plastics are traded globally, with significant flows of raw materials, intermediate products, and finished goods between countries and regions. Trade in plastics and plastic products contributes to economic interdependence, specialization, and comparative advantage, facilitating efficient allocation of resources and distribution of goods.

3. **Cost-effectiveness and Affordability:**
    - Plastics offer cost-effective solutions for manufacturing, packaging, and product design, enabling businesses to reduce production costs, enhance profitability, and offer competitive pricing to consumers. Plastic materials are lightweight, durable, and versatile, making them efficient and economical alternatives to traditional materials such as metal, glass, and wood.
    - The affordability and accessibility of plastic products have democratized consumption and improved living standards for billions of people worldwide. Plastics have become ubiquitous in everyday life, with applications ranging from food packaging and consumer goods to healthcare products and industrial components.
4. **Supply Chain Resilience and Reliability:**
    - The plastic industry plays a critical role in supply chain resilience and reliability, providing essential materials and components for manufacturing and production processes. Plastics offer properties such as durability, flexibility, and corrosion resistance, making them ideal for use in demanding industrial environments and harsh operating conditions.
    - Plastics contribute to the resilience of supply chains by reducing reliance on fragile or perishable materials, minimizing transportation costs, and optimizing inventory management. Plastic packaging, in particular, helps to protect goods from damage, spoilage, and contamination during storage, transportation, and distribution.
5. **Resource Efficiency and Waste Management:**
    - The plastic industry faces growing scrutiny and pressure to address environmental concerns related to plastic pollution, resource depletion, and waste management. Regulatory measures, voluntary initiatives, and market-based mechanisms are being implemented to promote resource efficiency, circularity, and sustainable practices.
    - Investments in recycling infrastructure, waste-to-energy technologies, and eco-friendly alternatives are driving innovation and diversification within the plastic industry. Circular economy principles, such as

extended producer responsibility (EPR) and product stewardship, are reshaping business models and supply chains, incentivizing producers to adopt more sustainable practices and reduce environmental impacts.

6. **Environmental Externalities and Costs:**
    - The plastic industry also incurs environmental externalities and costs associated with plastic pollution, resource extraction, and greenhouse gas emissions. The environmental impacts of plastic production and consumption, such as habitat destruction, wildlife mortality, and climate change, impose costs on society that are not reflected in market prices.
    - Efforts to internalize environmental externalities and address negative environmental impacts require policy interventions, market-based mechanisms, and collective action from stakeholders across the plastics value chain. Strategies such as pollution taxes, carbon pricing, and eco-labeling can help to incentivize sustainable behavior and promote environmental stewardship.

Chapter Four

Principles and Practices

Significance of ecosystem restoration

Ecosystem restoration refers to the process of repairing, rebuilding, or revitalizing degraded or damaged ecosystems to their original or natural state. It involves restoring the structure, function, and biodiversity of ecosystems through deliberate interventions aimed at reversing environmental degradation, enhancing ecosystem resilience, and promoting ecological health. Here's a detailed exploration of the definition and significance of ecosystem restoration:

1. **Definition**:
    - Ecosystem restoration encompasses a wide range of activities and techniques designed to rehabilitate ecosystems affected by human activities, natural disasters, or other disturbances. These activities may include habitat restoration, reforestation, wetland restoration, shoreline stabilization, invasive species control, and reintroduction of native species.
    - Ecosystem restoration aims to enhance the capacity of ecosystems to provide essential ecosystem services, such as clean air and water, carbon sequestration, soil fertility, pollination, and biodiversity conservation. It involves interdisciplinary approaches that integrate

ecological, social, economic, and cultural considerations to achieve sustainable outcomes.
2. **Significance**:
    o **Biodiversity Conservation**: Ecosystem restoration plays a crucial role in conserving biodiversity by creating or restoring habitats that support diverse plant and animal species. Healthy ecosystems provide food, shelter, and breeding grounds for wildlife, helping to maintain ecological balance and prevent species extinction. Restoring degraded habitats can help to conserve endangered species, protect biodiversity hotspots, and enhance ecosystem resilience to environmental changes.
    o **Climate Change Mitigation**: Ecosystem restoration contributes to climate change mitigation by sequestering carbon dioxide from the atmosphere and storing it in vegetation, soils, and biomass. Forests, wetlands, and mangroves are particularly effective at carbon sequestration, helping to offset greenhouse gas emissions and mitigate climate change impacts. Restoring degraded lands and ecosystems can also reduce vulnerability to climate-related disasters, such as floods, droughts, and wildfires.
    o **Natural Resource Management**: Ecosystem restoration promotes sustainable natural resource management by enhancing soil fertility, water quality, and nutrient cycling. Healthy ecosystems regulate hydrological processes, recharge groundwater aquifers, and prevent soil erosion, reducing the risk of floods, landslides, and desertification. Restoring degraded watersheds and riparian zones helps to protect water sources, enhance water security, and support agricultural productivity.
    o **Economic Benefits**: Ecosystem restoration generates a wide range of economic benefits for communities and societies, including job creation, income generation, and ecosystem services provisioning. Restored ecosystems provide opportunities for eco-tourism, recreation, and nature-based recreation, supporting local economies and livelihoods. Ecosystem services such as pollination, pest control,

and water purification have economic value and contribute to human well-being and quality of life.
- **Resilience and Adaptation**: Ecosystem restoration enhances ecosystem resilience and adaptive capacity to withstand environmental stresses, disturbances, and climate change impacts. Restored ecosystems are more resilient to droughts, storms, and extreme weather events, providing natural buffers against natural disasters and climate-related risks. Restoring coastal wetlands and mangroves helps to reduce coastal erosion, storm surge, and flooding, protecting communities and infrastructure.
- **Cultural and Societal Values**: Ecosystem restoration holds cultural and societal significance by preserving traditional knowledge, cultural heritage, and indigenous practices related to land stewardship and resource management. Restored ecosystems provide cultural ecosystem services such as spiritual, aesthetic, and recreational values, fostering connections between people and nature. Engaging local communities and indigenous peoples in restoration activities promotes social cohesion, empowerment, and environmental justice.

Successful restoration efforts

Successful ecosystem restoration efforts are guided by a set of principles and best practices informed by ecological science, adaptive management, and stakeholder engagement. These principles provide a framework for planning, implementing, and evaluating restoration projects to achieve ecological, social, and economic objectives. Here's a detailed exploration of the principles guiding successful restoration efforts:

1. **Clear Objectives and Targets**:
   - Restoration projects should have clear and measurable objectives, informed by ecological assessments, stakeholder input, and desired outcomes. Objectives may include restoring habitat connectivity, enhancing biodiversity, improving water quality, or mitigating erosion. Setting specific, achievable targets helps to guide decision-making, prioritize actions, and evaluate project success.
2. **Ecological Integrity and Functionality**:
   - Successful restoration efforts prioritize the restoration of ecological integrity and functionality by mimicking natural processes and ecosystem dynamics. This involves restoring key ecological functions such as nutrient cycling, water filtration, pollination, and seed dispersal. Using native species and habitats that are adapted to local environmental conditions enhances ecosystem resilience and promotes self-sustaining ecosystems over time.
3. **Site Selection and Planning**:
   - Site selection is critical for successful restoration, taking into account factors such as ecological suitability, landscape context, connectivity, and land tenure. Conducting thorough site assessments and feasibility studies helps to identify appropriate restoration sites, assess potential risks and constraints, and develop realistic restoration plans. Collaborating with landowners, stakeholders, and local communities ensures buy-in and support for restoration efforts.
4. **Adaptive Management and Monitoring**:

- Adaptive management involves iterative planning, implementation, monitoring, and adjustment based on feedback and new information. Monitoring key indicators such as species diversity, habitat structure, and ecosystem function helps to assess restoration progress, identify emerging issues, and inform adaptive management decisions. Flexibility and openness to learning from both successes and failures are essential for refining restoration strategies and improving outcomes over time.

5. **Stakeholder Engagement and Collaboration:**
    - Successful restoration efforts involve engaging stakeholders, including local communities, landowners, government agencies, NGOs, and indigenous peoples, throughout the restoration process. Meaningful participation and collaboration build trust, foster ownership, and promote shared stewardship of restored ecosystems. Incorporating traditional knowledge, cultural values, and local expertise enhances the relevance and effectiveness of restoration interventions.

6. **Long-term Commitment and Sustainability:**
    - Ecosystem restoration is a long-term endeavor that requires sustained commitment, funding, and management to achieve lasting results. Building institutional capacity, securing funding sources, and establishing legal frameworks and governance structures are essential for ensuring the sustainability of restoration initiatives. Integrating restoration into broader conservation strategies and land-use planning enhances its long-term viability and effectiveness.

7. **Ecosystem Services and Human Well-being:**
    - Successful restoration efforts consider the multiple benefits and ecosystem services provided by restored ecosystems, including clean water, air quality, climate regulation, and cultural values. Recognizing the interconnectedness between ecosystems and human well-being helps to justify investments in restoration and promote public support for conservation efforts. Restoring ecosystems that provide tangible benefits to local communities enhances social equity and improves livelihoods.

8. **Knowledge Sharing and Capacity Building**:
    - Sharing knowledge, best practices, and lessons learned from restoration projects facilitates learning, capacity building, and replication of successful approaches. Collaboration between scientists, practitioners, policymakers, and local communities fosters innovation, adaptive learning, and continuous improvement in restoration practice. Building technical skills, providing training opportunities, and disseminating information through outreach and education programs empower stakeholders to engage in restoration activities effectively.
9. **Resilience and Adaptation to Climate Change**:
    - Climate change presents challenges and opportunities for ecosystem restoration, requiring consideration of future climate projections, variability, and impacts. Designing restoration projects that enhance ecosystem resilience, promote species diversity, and buffer against climate-related risks helps ecosystems adapt to changing environmental conditions. Integrating climate-smart practices such as reforestation, green infrastructure, and natural flood management enhances the effectiveness and longevity of restoration efforts.
10. **Legislation, Policy Support, and Governance**:
    - Supportive legislation, policies, and governance mechanisms are essential for creating an enabling environment for ecosystem restoration. Governments, international agencies, and non-governmental organizations play critical roles in providing legal frameworks, regulatory incentives, and financial resources to support restoration initiatives. Aligning restoration efforts with national biodiversity targets, international agreements, and sustainable development goals strengthens their legitimacy and effectiveness.

Successful restoration projects

1. **Yamuna Biodiversity Park, India:**
   - The Yamuna Biodiversity Park in Delhi, India, is a notable example of successful urban ecosystem restoration. The park was established in 2002 on the floodplains of the Yamuna River, which had been heavily degraded due to urbanization, pollution, and encroachment.
   - The restoration project involved clearing invasive species, removing solid waste, and reintroducing native flora and fauna to recreate natural habitats such as wetlands, grasslands, and woodlands. The park now serves as a green lung for the city, providing recreational space, wildlife habitat, and educational opportunities for residents.
   - The Yamuna Biodiversity Park demonstrates the potential of urban restoration initiatives to enhance biodiversity, improve ecosystem services, and promote environmental awareness in densely populated areas.
2. **Loess Plateau Watershed Rehabilitation Project, China:**
   - The Loess Plateau Watershed Rehabilitation Project in China is one of the largest and most ambitious ecological restoration projects in the world. The project aims to combat soil erosion, desertification, and poverty in the Loess Plateau region, which had been severely degraded by centuries of intensive agriculture and deforestation.
   - The restoration project involves implementing soil and water conservation measures such as terracing, afforestation, and reforestation to stabilize soils, restore vegetation cover, and improve water retention. The project also includes sustainable land management practices, community engagement, and capacity building to support local livelihoods and enhance resilience to climate change.
   - The Loess Plateau Watershed Rehabilitation Project has successfully transformed degraded landscapes into productive, sustainable ecosystems, improving soil fertility, water quality, and biodiversity while

empowering local communities and improving their standard of living.

3. **Elwha River Ecosystem Restoration Project, USA:**
   - The Elwha River Ecosystem Restoration Project in Washington State, USA, is a landmark restoration initiative aimed at restoring natural river processes and habitats following the removal of two large hydroelectric dams. The Elwha and Glines Canyon Dams, built in the early 20th century, had blocked salmon migration, altered sediment transport, and degraded riparian habitats.
   - The restoration project involved the phased removal of the two dams between 2011 and 2014, allowing the river to regain its natural flow regime and reconnect spawning grounds in the upper river with the estuary and marine environment downstream. The restoration also included habitat enhancement measures such as revegetation, bank stabilization, and fish passage improvements.
   - The Elwha River Ecosystem Restoration Project has resulted in the recovery of salmon populations, restoration of riparian habitats, and revitalization of the river ecosystem. The project demonstrates the ecological and socio-economic benefits of dam removal and river restoration for aquatic ecosystems and local communities.

4. **Aral Sea Restoration Program, Central Asia:**
   - The Aral Sea Restoration Program in Central Asia is an international effort to address one of the world's most significant environmental disasters—the shrinking of the Aral Sea due to overuse of water for irrigation. The desiccation of the Aral Sea has led to ecological collapse, loss of biodiversity, and adverse impacts on human health and livelihoods.
   - The restoration program involves measures such as water conservation, reforestation, wetland restoration, and sustainable agriculture practices to improve water management and restore ecosystems in the Aral Sea basin. Efforts to increase water efficiency, restore wetlands, and rehabilitate degraded lands aim to stabilize the environment and mitigate the impacts of desertification and salinization.

- The Aral Sea Restoration Program demonstrates the importance of international cooperation, integrated water management, and sustainable development strategies in addressing complex environmental challenges and restoring degraded ecosystems on a regional scale.

5. **Great Green Wall Initiative, Africa:**
    - The Great Green Wall Initiative in Africa is a continent-wide effort to combat desertification, land degradation, and climate change by planting a wall of trees across the Sahel region of Africa. The initiative aims to restore degraded lands, improve soil fertility, and provide livelihood opportunities for millions of people living in the region.
    - The restoration project involves planting drought-resistant tree species such as acacias, baobabs, and indigenous shrubs along a 8,000-kilometer belt spanning from Senegal in the west to Djibouti in the east. The Great Green Wall initiative promotes community-led restoration efforts, sustainable land management practices, and climate-resilient agriculture to build resilience and adapt to changing environmental conditions.
    - The Great Green Wall Initiative demonstrates the potential of large-scale restoration projects to address land degradation, food insecurity, and poverty while providing climate change adaptation benefits for vulnerable communities in Africa.

Chapter Five

Challenges to Ecosystem Restoration

Political and economic obstacles

Political and economic obstacles pose significant challenges to ecosystem restoration efforts, affecting funding, policy formulation, stakeholder engagement, and project implementation. Here's a detailed exploration of the political and economic obstacles to ecosystem restoration:

1. **Lack of Political Will and Leadership:**
    - Ecosystem restoration requires strong political will and leadership to prioritize conservation, allocate resources, and enact supportive policies. However, political priorities often favor short-term economic gains over long-term environmental sustainability, leading to inadequate funding, weak enforcement of environmental regulations, and limited investment in restoration initiatives.
    - Political instability, corruption, and governance challenges can further undermine efforts to address environmental issues and implement restoration projects effectively. In some cases, vested interests and lobbying from powerful industries may influence decision-making and delay or obstruct environmental reforms.
2. **Competing Land Use and Development Pressures:**
    - Land use conflicts and development pressures present obstacles to ecosystem restoration by prioritizing economic activities such as agriculture, urbanization, infrastructure development, and extractive industries over conservation and restoration efforts. Rapid population growth, urban expansion, and land conversion for commercial

purposes encroach upon natural habitats, fragment ecosystems, and degrade biodiversity.
- Conflicting interests among stakeholders, including landowners, developers, conservationists, and indigenous communities, can lead to disputes over land rights, resource allocation, and environmental protection. Balancing competing demands for land and natural resources requires negotiation, compromise, and conflict resolution to reconcile economic development with environmental conservation goals.

3. **Limited Financial Resources and Funding Constraints**:
    - Ecosystem restoration projects often face funding constraints and resource limitations due to competing budget priorities, austerity measures, and economic downturns. Governments, international donors, and philanthropic organizations may allocate insufficient funding for restoration initiatives, hindering their scale, scope, and effectiveness.
    - Lack of dedicated funding mechanisms, financial incentives, and sustainable financing models for restoration projects can impede investment in long-term conservation and restoration efforts. The high upfront costs, uncertain returns on investment, and perceived risks associated with restoration projects may deter private sector involvement and hinder public-private partnerships.

4. **Policy and Regulatory Barriers**:
    - Policy and regulatory barriers, including outdated laws, bureaucratic hurdles, and inconsistent enforcement, can hinder ecosystem restoration by creating legal uncertainties, administrative delays, and compliance challenges. Regulatory frameworks may lack clear guidelines, standards, and incentives for restoration, making it difficult for stakeholders to navigate complex permitting processes and regulatory requirements.
    - Inadequate coordination and collaboration among government agencies, regulatory authorities, and stakeholders can exacerbate policy fragmentation, regulatory overlaps, and jurisdictional conflicts, undermining integrated planning and implementation of restoration projects. Streamlining

permitting procedures, harmonizing regulations, and enhancing inter-agency cooperation are essential for overcoming policy and regulatory barriers to ecosystem restoration.

5. **Short-term Economic Costs and Trade-offs:**
   - Ecosystem restoration often entails short-term economic costs, such as foregone revenues from land use conversion, lost agricultural productivity, and infrastructure investments required for restoration activities. These costs may outweigh the perceived benefits of restoration in the eyes of policymakers, landowners, and business interests, leading to reluctance to invest in restoration projects.
   - Failure to account for the long-term economic benefits and ecosystem services provided by restored ecosystems, such as carbon sequestration, water purification, and ecotourism, can result in undervaluation of restoration investments and missed opportunities for sustainable development. Cost-benefit analyses, economic valuation tools, and ecosystem service assessments can help to quantify the economic returns and social benefits of ecosystem restoration, making a stronger case for investment.

6. **Socio-political and Cultural Factors:**
   - Socio-political and cultural factors, including social inequalities, indigenous rights, and traditional land management practices, can influence ecosystem restoration outcomes and engagement with local communities. Inequitable distribution of costs and benefits, lack of consultation and participation, and cultural insensitivity can lead to social exclusion, resistance, and conflict, undermining the legitimacy and effectiveness of restoration efforts.
   - Recognizing and respecting the rights, knowledge, and cultural values of indigenous peoples and local communities is essential for building trust, fostering collaboration, and achieving socially inclusive restoration outcomes. Meaningful engagement, co-management arrangements, and equitable benefit-sharing mechanisms can empower communities to participate in decision-making and take ownership of restoration initiatives.

Technological limitations

Technological limitations can significantly impact the success and efficiency of ecosystem restoration efforts, affecting various aspects such as monitoring, assessment, implementation, and long-term management. Here's a detailed exploration of the technological limitations in ecosystem restoration:

1. **Remote Sensing and Monitoring**:
   - Remote sensing technologies, such as satellite imagery, LiDAR (Light Detection and Ranging), and aerial drones, are invaluable tools for monitoring ecosystem dynamics, assessing vegetation cover, and detecting changes in land use and land cover. However, limitations in spatial and temporal resolution, data accuracy, and cost can constrain their applicability for monitoring large-scale restoration projects.
   - Challenges such as cloud cover, sensor calibration, and data processing requirements can affect the reliability and timeliness of remote sensing data for monitoring restoration progress and assessing ecosystem health. Integration of multiple data sources, ground-truthing, and validation techniques are needed to overcome these limitations and improve the accuracy and reliability of remote sensing-based monitoring.
2. **GIS and Spatial Analysis**:
   - Geographic Information Systems (GIS) and spatial analysis tools are essential for spatial planning, site selection, and landscape-scale restoration planning. However, limitations in data availability, interoperability, and technical capacity can hinder the effective use of GIS for ecosystem restoration.
   - Lack of standardized protocols, data sharing agreements, and interoperable platforms can impede data integration, hindering decision-making and collaboration among stakeholders. Limited access to high-quality geospatial data, especially in developing countries and remote areas, further exacerbates disparities in technical capacity and resource

availability for restoration planning and implementation.
3. **Biotechnological Approaches**:
    - Biotechnological approaches, such as genetic engineering, tissue culture, and bioremediation, hold promise for enhancing the resilience and productivity of restored ecosystems. However, ethical concerns, regulatory barriers, and public perceptions about genetic modification and synthetic biology can hinder the adoption of biotechnological solutions for ecosystem restoration.
    - Uncertainties about the long-term ecological impacts, unintended consequences, and potential risks associated with genetically modified organisms (GMOs) and biotechnological interventions raise questions about their safety, efficacy, and sustainability in restoration practice. Stakeholder engagement, risk assessment, and regulatory oversight are essential for addressing these concerns and ensuring responsible deployment of biotechnological approaches.
4. **Innovations in Seed Technology and Propagation**:
    - Seed technology and propagation techniques play a crucial role in large-scale revegetation and habitat restoration projects. Advances in seed banking, seed storage, and seedling production have improved the availability and quality of plant materials for restoration purposes. However, challenges such as seed dormancy, low germination rates, and genetic diversity loss can limit the effectiveness of seed-based restoration efforts.
    - Invasive species, habitat degradation, and climate change can affect seed availability, viability, and germination success, posing challenges for sourcing and using native plant materials in restoration projects. Developing resilient seed sources, propagating locally adapted genotypes, and incorporating genetic diversity conservation strategies can enhance the resilience and success of restoration plantings.
5. **Soil Amendments and Bioengineering**:
    - Soil amendments and bioengineering techniques, such as biochar application, mycorrhizal inoculation,

and soil stabilization measures, are used to improve soil fertility, structure, and stability in degraded ecosystems. However, limitations in scalability, cost-effectiveness, and ecological compatibility can constrain their application in large-scale restoration projects.
- Bioengineering techniques, such as erosion control blankets, vegetative barriers, and biodegradable mulches, can help stabilize slopes, reduce soil erosion, and enhance revegetation success. However, challenges such as installation costs, maintenance requirements, and limited effectiveness in extreme environments can limit their applicability and long-term durability in restoration practice.

6. **Data Management and Integration**:
    - Effective data management and integration are essential for synthesizing diverse sources of information, sharing knowledge, and informing decision-making in ecosystem restoration. However, challenges such as data silos, format compatibility, and data quality control can hinder data interoperability, hindering collaboration and hindering the implementation of integrated restoration strategies.
    - Developing standardized data formats, metadata standards, and data sharing protocols can facilitate data integration, improve data accessibility, and promote transparency in restoration projects. Leveraging emerging technologies such as cloud computing, big data analytics, and machine learning can enhance data processing capabilities and support evidence-based decision-making in ecosystem restoration.

7. **Capacity Building and Training**:
    - Capacity building and training are essential for building technical skills, enhancing institutional capacity, and empowering stakeholders to engage effectively in ecosystem restoration. However, limitations in access to training resources, technical expertise, and institutional support can hinder efforts to develop local capacity and empower communities to participate in restoration activities.

- Inadequate investment in education, training programs, and professional development opportunities for restoration practitioners, land managers, and community members can perpetuate disparities in technical capacity and hinder the adoption of best practices and innovative approaches in restoration practice. Investing in capacity-building initiatives, knowledge exchange platforms, and experiential learning opportunities can enhance the effectiveness and sustainability of ecosystem restoration efforts.

Social and cultural barriers

Social and cultural barriers can significantly impede ecosystem restoration efforts by influencing stakeholder perceptions, community engagement, and participation in conservation initiatives. These barriers stem from cultural norms, socioeconomic disparities, historical legacies, and power dynamics that shape attitudes and behaviors towards the environment. Here's a detailed exploration of the social and cultural barriers to ecosystem restoration:

1. **Limited Awareness and Environmental Education**:
    - Lack of awareness and environmental education about the importance of biodiversity, ecosystem services, and the benefits of restoration can hinder community support and engagement in restoration efforts. Many people may not fully understand the ecological functions of ecosystems or the consequences of environmental degradation, leading to apathy or indifference towards conservation.
    - Inadequate access to environmental education, outreach programs, and awareness campaigns, particularly in marginalized or underserved communities, exacerbates disparities in environmental literacy and awareness. Addressing knowledge gaps and promoting environmental education from an early age can help overcome this barrier and foster a culture of environmental stewardship and conservation.
2. **Cultural Beliefs and Traditional Practices**:
    - Cultural beliefs, values, and traditional practices may conflict with conservation goals and restoration activities, leading to resistance or opposition from local communities. Cultural norms related to land use, resource extraction, and livelihood activities

may prioritize short-term economic gains over long-term environmental sustainability, hindering efforts to conserve and restore ecosystems.
- Indigenous peoples and local communities may have deep spiritual, cultural, and historical connections to their land and natural resources, influencing their attitudes towards conservation and restoration initiatives. Respecting indigenous knowledge, traditional land management practices, and customary rights is essential for building trust, fostering collaboration, and achieving mutually beneficial outcomes in restoration projects.

3. **Socioeconomic Inequities and Livelihood Dependence**:
    - Socioeconomic disparities and livelihood dependence on natural resources can create barriers to ecosystem restoration by prioritizing immediate economic needs over long-term environmental considerations. Communities reliant on ecosystem services for food, water, fuel, and livelihoods may perceive restoration efforts as threats to their economic security or cultural identity.
    - Poverty, unemployment, and lack of alternative livelihood opportunities can exacerbate dependency on natural resources and drive unsustainable land use practices, such as deforestation, overgrazing, and illegal logging. Addressing underlying socio-economic drivers, promoting sustainable livelihoods, and providing incentives for conservation can help overcome barriers related to economic dependence and insecurity.

4. **Power Dynamics and Governance Challenges**:
    - Power dynamics and governance challenges, including unequal access to resources, decision-making processes, and representation in decision-making bodies, can marginalize certain groups and limit their participation in restoration initiatives. Women, indigenous peoples, and marginalized communities are often disproportionately affected by environmental degradation and may face barriers to accessing information, resources, and decision-making opportunities.
    - Top-down approaches to conservation and restoration that fail to incorporate local perspectives,

traditional knowledge, and community priorities can lead to resistance, mistrust, and conflicts with stakeholders. Empowering marginalized groups, promoting inclusive governance structures, and fostering participatory decision-making processes are essential for overcoming power imbalances and ensuring equitable engagement in restoration efforts.

5. **Land Tenure and Property Rights**:
   o Land tenure insecurity, unclear property rights, and competing land claims can create obstacles to ecosystem restoration by limiting land access, investment, and tenure security for restoration projects. Ambiguous land tenure arrangements, unresolved land conflicts, and competing land uses can hinder efforts to secure land tenure for restoration purposes.
   o Lack of legal recognition of indigenous land rights, customary land tenure systems, and community-based resource management regimes can undermine local ownership, control, and stewardship of natural resources. Strengthening land tenure rights, promoting land titling programs, and resolving land disputes through participatory mechanisms can facilitate collaboration and partnership in restoration initiatives.

6. **Language and Communication Barriers**:
   o Language barriers, cultural differences, and communication gaps between restoration practitioners and local communities can hinder effective engagement, collaboration, and knowledge sharing in restoration projects. Misunderstandings, misinterpretations, and mistranslations of technical terms and concepts can lead to confusion, distrust, and disengagement among stakeholders.
   o Tailoring communication strategies, using culturally appropriate language, and employing participatory approaches such as community-based mapping, storytelling, and visual aids can improve communication and facilitate meaningful dialogue between diverse stakeholders. Building trust, fostering reciprocal relationships, and promoting cultural sensitivity are essential for bridging

communication barriers and promoting inclusive participation in restoration efforts.
7. **Resistance to Change and Risk Aversion:**
    o Resistance to change and risk aversion among stakeholders can impede ecosystem restoration efforts by perpetuating inertia, maintaining the status quo, and resisting innovation or experimentation. Fear of failure, uncertainty about outcomes, and reluctance to adopt new approaches or technologies can hinder adaptive management and inhibit learning from experience.
    o Overcoming resistance to change requires fostering a culture of experimentation, learning, and adaptation, where failures are viewed as opportunities for improvement and innovation. Building capacity for adaptive management, promoting learning networks, and providing incentives for experimentation can help overcome barriers related to risk aversion and resistance to change.

Chapter Six

Innovative Solutions to End Plastic Pollution

Biodegradable alternatives to traditional plastics

Biodegradable alternatives to traditional plastics offer promising solutions to mitigate the environmental impact of plastic pollution. These alternatives are designed to degrade naturally in the environment, reducing the accumulation of persistent plastic waste in landfills, oceans, and ecosystems. Here's a detailed exploration of biodegradable alternatives to traditional plastics:

1. **Biodegradable Polymers:**
    - Biodegradable polymers are materials derived from renewable resources or synthesized from biologically derived monomers that can break down into natural compounds under specific environmental conditions. These polymers include:
        - **Polylactic Acid (PLA):** PLA is derived from renewable resources such as corn starch, sugarcane, or cassava. It is compostable and biodegradable under industrial composting conditions, where high temperatures and microbial activity facilitate degradation. PLA can be used in various applications, including packaging, disposable utensils, and food containers.
        - **Polyhydroxyalkanoates (PHA):** PHAs are biodegradable polymers produced by microbial fermentation of renewable feedstocks such as sugars or plant oils. They

are fully biodegradable in various environments, including soil, marine, and anaerobic conditions. PHAs exhibit properties similar to conventional plastics and can be used in packaging, agricultural films, and biomedical applications.
- **Polybutylene Succinate (PBS)**: PBS is a biodegradable polyester derived from renewable resources such as succinic acid and 1,4-butanediol. It is compostable and biodegradable under aerobic and anaerobic conditions, making it suitable for applications such as mulch films, agricultural products, and disposable packaging.
- **Polyhydroxybutyrate (PHB)**: PHB is a biodegradable polyester produced by bacterial fermentation of renewable carbon sources. It is fully biodegradable in various environments and exhibits properties similar to conventional plastics. PHB can be used in applications such as packaging, biomedical devices, and consumer products.

2. **Biodegradable Additives**:
    - Biodegradable additives are compounds added to conventional plastics to enhance their biodegradability and accelerate degradation in the environment. These additives include:
        - **Oxo-degradable Additives**: Oxo-degradable additives accelerate the fragmentation and degradation of plastics through oxidation when exposed to heat, light, or mechanical stress. They break down plastics into smaller fragments that are more susceptible to microbial degradation in the environment.
        - **Biodegradable Fillers**: Biodegradable fillers such as starch, cellulose, or lignin can be incorporated into conventional plastics to enhance their biodegradability and reduce their environmental impact. These fillers provide reinforcement, improve mechanical properties, and facilitate degradation by microbes in the environment.

- **Pro-oxidant Additives**: Pro-oxidant additives initiate oxidative degradation of plastics, leading to chain scission and fragmentation of polymer molecules. They promote the breakdown of plastics into smaller fragments that are more easily biodegraded by microorganisms in the environment.

3. **Bio-based Plastics**:
   - Bio-based plastics are derived from renewable biomass sources such as plants, algae, or microorganisms. These plastics can be biodegradable or non-biodegradable, depending on their chemical composition and processing methods. Bio-based plastics include:
     - **Bio-based Polyethylene (Bio-PE)**: Bio-PE is derived from renewable feedstocks such as sugarcane ethanol or vegetable oils. It has similar properties to conventional polyethylene and can be used in various applications, including packaging, containers, and films. Bio-PE can be biodegradable or non-biodegradable, depending on its composition.
     - **Bio-based Polyethylene Terephthalate (Bio-PET)**: Bio-PET is derived from renewable feedstocks such as sugarcane or corn starch. It is chemically identical to conventional PET and can be recycled using existing PET recycling infrastructure. Bio-PET can be used in applications such as beverage bottles, food packaging, and textiles.
     - **Bio-based Polyamides (Bio-PA)**: Bio-PA is derived from renewable sources such as castor oil or bio-based monomers. It exhibits similar properties to conventional polyamides and can be used in applications such as automotive parts, engineering plastics, and consumer goods.

4. **Compostable Plastics**:
   - Compostable plastics are designed to biodegrade under controlled composting conditions, where high temperatures, humidity, and microbial activity

facilitate degradation into organic matter, water, and carbon dioxide. Compostable plastics include:

- **Compostable Polymers**: Compostable polymers such as PLA, PHA, or starch-based plastics are certified to biodegrade in industrial composting facilities within specific timeframes and under controlled conditions. They can be used in applications such as food service ware, compostable bags, and agricultural mulches.
- **Biodegradable Packaging**: Compostable packaging materials such as compostable films, bags, and containers are designed to replace conventional plastic packaging and reduce waste in landfills. These materials are certified compostable and meet stringent standards for biodegradability, compostability, and eco-toxicity.

5. **Challenges and Considerations**:
   - Despite their potential environmental benefits, biodegradable alternatives to traditional plastics face challenges such as:
     - **Cost**: Biodegradable plastics often have higher production costs compared to conventional plastics, limiting their widespread adoption and market penetration.
     - **Performance**: Biodegradable plastics may exhibit inferior mechanical properties, durability, and shelf-life compared to conventional plastics, limiting their suitability for certain applications.
     - **End-of-Life Management**: Biodegradable plastics require specific end-of-life management systems, such as industrial composting facilities or anaerobic digestion plants, to biodegrade effectively. Inadequate infrastructure and lack of composting facilities can hinder their disposal and recycling.
     - **Certification and Standards**: Biodegradable plastics must meet stringent certification and

standards for biodegradability, compostability, and eco-toxicity to ensure their environmental performance and safety. Lack of standardized testing methods and certification schemes can lead to confusion and greenwashing in the marketplace.

Technological advancements in waste management

Technological advancements in waste management play a crucial role in addressing the growing challenges of waste generation, disposal, and environmental pollution. These advancements encompass a wide range of innovative technologies and approaches aimed at improving waste collection, sorting, recycling, treatment, and disposal processes. Here's a detailed exploration of technological advancements in waste management:

1. **Smart Waste Collection Systems**:
    - Smart waste collection systems leverage technology such as sensors, IoT (Internet of Things), and data analytics to optimize waste collection routes, schedules, and resource allocation. These systems use real-time data on fill levels, collection frequency, and route efficiency to optimize collection operations, reduce fuel consumption, and minimize environmental impact.
    - Sensor-equipped waste bins and containers detect fill levels and transmit data to centralized management platforms, enabling route optimization, predictive maintenance, and remote monitoring of waste collection activities. Smart waste collection systems improve operational efficiency, reduce costs, and enhance service quality by providing timely and data-driven decision-making.
2. **Automated Sorting and Recycling Technologies**:
    - Automated sorting and recycling technologies utilize robotics, AI (Artificial Intelligence), and machine learning algorithms to enhance the efficiency and accuracy of waste sorting and recycling processes. These technologies enable high-speed sorting of recyclable materials, such as plastics, metals, glass, and paper, based on their properties, composition, and recyclability.
    - Optical sorting systems use sensors and cameras to identify and separate different types of recyclable materials from mixed waste streams. Robotics and AI-powered sorting robots further automate the sorting process by precisely picking and segregating materials based on predefined criteria. These

technologies increase recycling rates, reduce contamination, and improve the quality of recycled materials.
3. **Advanced Waste-to-Energy Technologies**:
    - Waste-to-energy technologies convert organic waste materials into renewable energy sources, such as electricity, heat, or biofuels, through processes such as incineration, gasification, anaerobic digestion, and pyrolysis. These technologies help reduce landfilling, mitigate greenhouse gas emissions, and recover valuable energy resources from waste.
    - Anaerobic digestion facilities use microorganisms to break down organic waste in the absence of oxygen, producing biogas (methane) and digestate as byproducts. Biogas can be used for electricity generation, heating, or vehicle fuel, while digestate can be used as fertilizer or soil conditioner. Advanced gasification and pyrolysis technologies convert organic waste into syngas or bio-oil, which can be used as feedstock for renewable energy production or chemical synthesis.
4. **Chemical Recycling and Upcycling Processes**:
    - Chemical recycling and upcycling processes convert waste plastics into new raw materials or higher-value products through depolymerization, chemical conversion, or catalytic upgrading. These processes complement mechanical recycling by enabling the recycling of mixed, contaminated, or difficult-to-recycle plastics that are not suitable for traditional recycling methods.
    - Depolymerization technologies break down plastic polymers into monomers or smaller molecular compounds, which can be used to produce new plastics or chemical feedstocks. Upcycling technologies transform waste plastics into value-added products, such as fuels, chemicals, construction materials, or 3D printing filaments, with enhanced properties or performance.
5. **Landfill Management and Remediation Technologies**:
    - Landfill management and remediation technologies aim to minimize the environmental impact of landfills, mitigate leachate and methane emissions,

and reclaim land for reuse or redevelopment. These technologies include:

- Landfill gas recovery systems capture and treat methane emissions from decomposing organic waste in landfills, reducing greenhouse gas emissions and generating renewable energy. Gas collection systems extract methane-rich landfill gas for electricity generation, heating, or direct use as a fuel.
- Leachate treatment systems remove contaminants from landfill leachate, a liquid byproduct formed when rainwater percolates through waste materials, to prevent groundwater contamination and surface water pollution. Advanced treatment technologies such as membrane filtration, activated carbon adsorption, and biological oxidation remove pollutants and improve effluent quality.
- Landfill closure and remediation technologies restore decommissioned landfills to their natural state or repurpose them for sustainable land use, such as parks, green spaces, or renewable energy facilities. Remediation techniques such as landfill capping, soil stabilization, and phytoremediation mitigate environmental risks and protect human health.

6. **Waste Minimization and Circular Economy Solutions**:
    - Waste minimization and circular economy solutions focus on reducing waste generation, promoting resource efficiency, and maximizing the reuse, recycling, and recovery of materials and products. These solutions include:

        - Product redesign and eco-design strategies aim to minimize the environmental impact of products throughout their lifecycle by optimizing material use, reducing packaging, and enhancing durability, reparability, and recyclability.
        - Extended producer responsibility (EPR) schemes shift responsibility for managing

end-of-life products and packaging from consumers to producers, encouraging manufacturers to design products for reuse, recycling, or recovery and finance the collection, recycling, and disposal of their products.
- Closed-loop recycling systems aim to close the material loop by recovering and reintegrating recycled materials back into the production process, reducing the demand for virgin resources and minimizing waste generation. These systems promote circularity, resource conservation, and sustainable consumption and production patterns.

Policy recommendations for reducing plastic usage

Policy recommendations for reducing plastic usage are essential for addressing the environmental and societal challenges associated with plastic pollution. Effective policies can help incentivize behavior change, promote sustainable alternatives, and foster a transition towards a circular economy. Here's a detailed exploration of policy recommendations for reducing plastic usage:

1. **Single-Use Plastic Bans and Restrictions**:
   - Implement comprehensive bans or restrictions on single-use plastics, such as plastic bags, straws, cutlery, and disposable packaging, to reduce plastic consumption and encourage the adoption of reusable alternatives.
   - Enforce regulations to prohibit the production, sale, and distribution of certain types of single-use plastics, while promoting eco-friendly alternatives such as reusable bags, stainless steel straws, and compostable packaging.
2. **Extended Producer Responsibility (EPR) Regulations**:
   - Establish EPR regulations that hold producers responsible for the end-of-life management of their products, including collection, recycling, and disposal. Require producers to design products for durability, reparability, and recyclability, and finance the collection and recycling infrastructure necessary to manage their products' waste.
   - Implement deposit-return schemes (DRS) for beverage containers to incentivize recycling and reduce littering. Consumers pay a deposit fee when purchasing beverages in containers, which is refunded when they return the empty containers for recycling.
3. **Plastic Packaging Regulations**:
   - Introduce regulations to reduce the use of plastic packaging and promote sustainable packaging alternatives, such as paper-based packaging, biodegradable materials, and reusable packaging systems.
   - Mandate labeling requirements for plastic packaging to inform consumers about the environmental impact

of different packaging materials, recycling instructions, and eco-friendly alternatives.
4. **Taxation and Pricing Mechanisms**:
    o Impose taxes or levies on plastic products, packaging, or virgin plastic materials to internalize the environmental costs of plastic pollution and incentivize the use of sustainable alternatives.
    o Implement differential pricing strategies that make plastic products more expensive relative to environmentally friendly alternatives, encouraging consumers to make sustainable purchasing choices.
5. **Plastic Bag Fees and Charges**:
    o Implement plastic bag fees or charges to discourage the use of single-use plastic bags and promote the adoption of reusable bags. Revenues generated from bag fees can be reinvested in waste management infrastructure, environmental education, or conservation programs.
    o Provide exemptions or subsidies for low-income households to ensure equitable access to reusable bags and minimize the regressive impact of plastic bag fees on vulnerable populations.
6. **Public Awareness and Education Campaigns**:
    o Launch public awareness and education campaigns to raise awareness about the environmental impact of plastic pollution, promote sustainable consumption habits, and encourage behavior change.
    o Provide information and resources to consumers about alternatives to single-use plastics, recycling best practices, and the benefits of reducing plastic waste for the environment and human health.
7. **Incentives for Innovation and Research**:
    o Provide financial incentives, grants, and research funding to support the development of innovative solutions for plastic waste reduction, recycling technologies, and sustainable packaging alternatives.
    o Foster collaboration between government agencies, academic institutions, research organizations, and private sector stakeholders to accelerate the transition towards a circular economy and develop scalable solutions to plastic pollution.
8. **International Cooperation and Agreements**:

- Collaborate with other countries and international organizations to develop global agreements, protocols, and standards for plastic pollution reduction, waste management, and marine conservation.
- Participate in regional initiatives and partnerships to address transboundary plastic pollution, promote sustainable consumption and production practices, and strengthen waste management infrastructure in developing countries.

9. **Policy Integration and Coordination**:
   - Integrate plastic pollution reduction policies with broader environmental, climate, and sustainable development goals to maximize synergies and minimize trade-offs.
   - Establish intergovernmental task forces, multi-stakeholder platforms, and coordination mechanisms to facilitate policy coherence, knowledge sharing, and best practices exchange on plastic pollution reduction and waste management.

10. **Monitoring, Evaluation, and Enforcement**:
    - Establish monitoring and evaluation mechanisms to assess the effectiveness and impact of plastic pollution reduction policies, track progress towards targets, and identify areas for improvement.
    - Strengthen enforcement mechanisms, penalties, and compliance measures to ensure regulatory compliance and deter illegal activities such as plastic dumping, littering, and illegal waste trade.

Chapter Seven

Collaborative Efforts and Community Engagement

Importance of interdisciplinary collaboration

Interdisciplinary collaboration plays a pivotal role in addressing complex challenges and driving innovation across various fields, including science, technology, healthcare, environmental conservation, and social policy. By bringing together experts from diverse disciplines, interdisciplinary collaboration fosters creativity, facilitates knowledge exchange, and generates holistic solutions to multifaceted problems. Here's a detailed exploration of the importance of interdisciplinary collaboration:

1. **Holistic Problem-Solving**:
    - Interdisciplinary collaboration enables stakeholders from different disciplines to combine their expertise, perspectives, and methodologies to tackle complex problems from multiple angles. By integrating diverse viewpoints and approaches, interdisciplinary teams can develop comprehensive solutions that consider the interconnectedness of social, environmental, economic, and technical factors.
2. **Innovation and Creativity**:
    - Interdisciplinary collaboration fosters innovation by breaking down disciplinary silos, fostering cross-pollination of ideas, and promoting unconventional thinking. By bringing together individuals with diverse backgrounds, skills, and experiences, interdisciplinary teams can spark creativity, inspire new approaches, and generate breakthrough solutions that may not be possible within the confines of a single discipline.
3. **Transdisciplinary Insights**:
    - Interdisciplinary collaboration goes beyond multidisciplinary approaches by transcending disciplinary boundaries and integrating insights from

diverse fields into a unified framework. Transdisciplinary research and collaboration involve co-creating knowledge, synthesizing different perspectives, and addressing real-world problems that require interdisciplinary solutions.
4. **Addressing Complex Challenges**:
    o Many of the most pressing challenges facing society today, such as climate change, public health crises, urbanization, and poverty, are inherently complex and multifaceted. Interdisciplinary collaboration is essential for understanding the root causes of these challenges, identifying interconnected drivers, and developing integrated strategies for mitigation and adaptation.
5. **Cross-Sectoral Partnerships**:
    o Interdisciplinary collaboration fosters partnerships and collaborations across sectors, including academia, government, industry, non-profit organizations, and civil society. These cross-sectoral partnerships leverage complementary strengths, resources, and networks to address shared challenges and achieve collective impact.
6. **Translational Research and Innovation**:
    o Interdisciplinary collaboration accelerates the translation of research findings and technological innovations into real-world applications and solutions. By bridging the gap between academic research and practical implementation, interdisciplinary teams can facilitate technology transfer, commercialization, and uptake of innovations in diverse sectors.
7. **Policy and Decision-Making**:
    o Interdisciplinary collaboration informs evidence-based policy and decision-making by providing policymakers with a holistic understanding of complex issues, diverse perspectives, and potential trade-offs. By integrating scientific evidence, socio-economic analysis, and stakeholder engagement, interdisciplinary research can inform policy development, implementation, and evaluation processes.

8. **Education and Capacity Building:**
    - Interdisciplinary collaboration enriches educational experiences and promotes lifelong learning by exposing students, researchers, and practitioners to diverse perspectives, methodologies, and problem-solving approaches. Interdisciplinary education and training programs prepare future leaders to navigate complex challenges, work effectively in interdisciplinary teams, and communicate across disciplinary boundaries.
9. **Resilience and Adaptation:**
    - Interdisciplinary collaboration enhances resilience and adaptive capacity by fostering flexibility, innovation, and learning in the face of uncertainty and change. By integrating diverse perspectives and expertise, interdisciplinary teams can anticipate emerging challenges, identify opportunities for adaptation, and develop flexible strategies for resilience building.
10. **Promoting Equity and Inclusion:**
    - Interdisciplinary collaboration promotes equity and inclusion by facilitating participation and representation from diverse communities, cultures, and backgrounds. By valuing diverse perspectives and promoting inclusivity, interdisciplinary teams can address systemic inequalities, challenge dominant paradigms, and promote social justice in research, policy, and practice.

Engaging communities in restoration efforts

Engaging communities in restoration efforts is essential for achieving long-term success, sustainability, and social acceptance of restoration projects. Community engagement fosters local ownership, builds social capital, and enhances the resilience of restored ecosystems by incorporating traditional knowledge, values, and cultural practices. Here's a detailed exploration of strategies for engaging communities in restoration efforts:

1. **Stakeholder Identification and Mapping**:
    - Identify and map key stakeholders, including local communities, indigenous peoples, government agencies, NGOs, businesses, and landowners, who are affected by or have an interest in restoration activities.
    - Conduct stakeholder analyses to assess their interests, concerns, priorities, and potential contributions to restoration efforts, and tailor engagement strategies accordingly.
2. **Participatory Planning and Decision-Making**:
    - Facilitate participatory planning processes that involve communities in setting restoration goals, identifying priority areas, and designing restoration interventions that align with local needs, aspirations, and cultural values.
    - Organize community meetings, workshops, focus groups, and participatory mapping exercises to solicit input, gather feedback, and co-design restoration plans that reflect diverse perspectives and preferences.
3. **Capacity Building and Training**:
    - Build local capacity and empower communities to participate actively in restoration activities through training programs, workshops, and skill-building initiatives.
    - Provide technical assistance, knowledge exchange opportunities, and resources to enhance community members' understanding of restoration techniques, ecological principles, and sustainable land management practices.

4. **Cultural Sensitivity and Respect:**
    o Respect and incorporate traditional knowledge, cultural practices, and indigenous wisdom into restoration planning and implementation processes.
    o Consult with indigenous elders, community leaders, and cultural experts to ensure that restoration activities are culturally appropriate, respectful of sacred sites, and aligned with traditional land stewardship practices.
5. **Communication and Outreach:**
    o Establish clear and transparent communication channels to disseminate information, share updates, and engage with community members throughout the restoration process.
    o Use a variety of communication tools and mediums, including community meetings, radio broadcasts, social media, newsletters, and signage, to reach diverse audiences and facilitate two-way dialogue.
6. **Inclusive Participation and Representation:**
    o Ensure inclusive participation and representation of diverse stakeholders, including marginalized groups, women, youth, and vulnerable populations, in restoration decision-making processes.
    o Create inclusive spaces for dialogue, deliberation, and collaboration, where all voices are heard, respected, and valued, and decision-making processes are transparent and accountable.
7. **Local Employment and Economic Opportunities:**
    o Create local employment opportunities and economic incentives for community members to participate in restoration activities, such as tree planting, habitat restoration, and eco-tourism initiatives.
    o Support the development of sustainable livelihoods, income-generating activities, and small-scale enterprises that are compatible with restoration objectives and contribute to local economic development.
8. **Community-Based Monitoring and Evaluation:**
    o Engage communities in monitoring and evaluating restoration progress, outcomes, and impacts by involving them in data collection, observation, and reporting activities.

- Establish community-based monitoring systems, citizen science initiatives, and participatory research projects that empower communities to assess the effectiveness of restoration interventions, track changes in ecosystem health, and provide feedback for adaptive management.

9. **Recognition and Incentives**:
    - Acknowledge and recognize the contributions of community members, volunteers, and local organizations to restoration efforts through public recognition, awards, and incentives.
    - Provide tangible benefits, such as access to restored natural areas, recreational opportunities, or revenue-sharing arrangements, to incentivize community participation and stewardship of restored ecosystems.

10. **Long-Term Engagement and Partnership Building**:
    - Foster long-term relationships, trust, and collaboration with communities by building meaningful partnerships, maintaining ongoing dialogue, and demonstrating commitment to shared goals and values.
    - Invest in building social capital, fostering reciprocity, and nurturing a sense of ownership and pride in restored landscapes, which can lead to sustained community engagement and stewardship over time.

Grassroots initiatives and citizen science

Grassroots initiatives and citizen science play a crucial role in environmental conservation, biodiversity monitoring, and ecosystem restoration efforts. These bottom-up approaches empower local communities, volunteers, and citizen scientists to actively participate in scientific research, data collection, and community-based conservation projects. Here's a detailed exploration of grassroots initiatives and citizen science:

1. **Definition and Principles**:
   - Grassroots initiatives are community-led, decentralized efforts that emerge from the local level to address specific social, environmental, or economic challenges. They are characterized by bottom-up decision-making, volunteer participation, and collective action aimed at achieving positive change.
   - Citizen science involves the participation of non-professional scientists, volunteers, and community members in scientific research, data collection, and monitoring activities. It engages citizens in the scientific process, promotes public awareness, and generates valuable data for scientific inquiry and conservation efforts.
2. **Community Empowerment and Participation**:
   - Grassroots initiatives empower local communities to take ownership of environmental issues, advocate for change, and implement solutions that address their needs and priorities. They provide a platform for community members to voice their concerns, share knowledge, and mobilize resources for collective action.
   - Citizen science engages volunteers in scientific research and monitoring activities, enabling them to contribute to biodiversity conservation, habitat restoration, and environmental monitoring efforts. By involving citizens in data collection, analysis, and interpretation, citizen science projects foster a sense of ownership, stewardship, and connection to the natural world.

3. **Data Collection and Monitoring**:
   - Grassroots initiatives and citizen science projects collect valuable data on biodiversity, ecosystem health, and environmental quality that contribute to scientific understanding, decision-making, and policy development. Volunteers and citizen scientists participate in field surveys, data recording, and monitoring activities, providing researchers with large datasets and spatial coverage that complement traditional scientific monitoring efforts.
   - Grassroots monitoring programs often focus on specific species, habitats, or ecosystems of local significance, such as urban green spaces, community gardens, or riparian zones. Citizen science projects may involve monitoring bird populations, tracking wildlife movements, conducting water quality assessments, or documenting phenological changes in plant communities.
4. **Environmental Education and Awareness**:
   - Grassroots initiatives and citizen science projects raise public awareness about environmental issues, foster environmental literacy, and promote active citizenship. They provide opportunities for experiential learning, hands-on engagement, and nature-based education that connect people with their local environment and inspire environmental stewardship.
   - Citizen science projects often incorporate educational components, such as training workshops, field trips, and outreach events, to build participants' scientific skills, enhance their understanding of ecological concepts, and cultivate a sense of environmental responsibility.
5. **Community-Based Conservation and Restoration**:
   - Grassroots initiatives mobilize communities to undertake conservation and restoration activities at the local level, such as habitat restoration, invasive species removal, and native plant propagation. They leverage local knowledge, resources, and social networks to implement on-the-ground actions that enhance biodiversity, improve ecosystem resilience, and promote sustainable land management practices.

- Citizen science projects contribute to conservation efforts by monitoring threatened species, identifying invasive species, and assessing the effectiveness of restoration interventions. Volunteers and citizen scientists play a critical role in collecting data on species distributions, population trends, and habitat quality, which inform conservation planning and management decisions.
6. **Policy Advocacy and Social Change**:
    - Grassroots initiatives and citizen science projects advocate for policy change, community-based management approaches, and sustainable development practices that promote environmental conservation and social equity. They engage in lobbying, public campaigns, and grassroots organizing to raise awareness, influence decision-makers, and mobilize support for conservation initiatives.
    - Citizen science data and findings contribute to evidence-based policymaking, environmental monitoring, and resource management decisions at local, regional, and national levels. By providing policymakers with locally relevant data and community perspectives, citizen science projects empower citizens to participate in governance processes and shape environmental policies that reflect their needs and priorities.
7. **Examples of Grassroots Initiatives and Citizen Science Projects**:
    - **Community-based monitoring of water quality**: Local communities monitor rivers, lakes, and streams for pollution, nutrient runoff, and aquatic habitat degradation using simple monitoring tools and protocols.
    - **Citizen science bird counts**: Volunteers participate in annual bird counts, such as the Audubon Christmas Bird Count or the Great Backyard Bird Count, to track bird populations and migration patterns.
    - **Community-led restoration projects**: Neighborhood groups collaborate to restore degraded habitats, plant native trees, and create green spaces in urban areas to improve biodiversity and ecosystem services.

- **Invasive species monitoring and management**: Citizen scientists help identify and map invasive plant and animal species, such as European starlings or Japanese knotweed, and assist in removal efforts to protect native ecosystems.

Chapter Eight

The Future of Plastic-Free Ecosystems

Vision for a plastic-free future

A vision for a plastic-free future is characterized by a paradigm shift towards sustainable consumption and production patterns that prioritize environmental protection, human health, and social well-being. It entails a transition away from the pervasive use of single-use plastics, non-recyclable materials, and wasteful packaging towards a circular economy model where resources are conserved, reused, and recycled in closed-loop systems. Here's a detailed exploration of the vision for a plastic-free future:

1. **Reduction of Single-Use Plastics**:
    - In a plastic-free future, single-use plastics such as plastic bags, straws, bottles, and packaging are significantly reduced or eliminated altogether through policy interventions, consumer behavior change, and industry innovation.
    - Governments implement comprehensive bans, restrictions, and regulations on single-use plastics, incentivize the use of reusable alternatives, and promote circular economy principles to minimize plastic waste generation.
2. **Transition to Sustainable Materials**:
    - The use of sustainable materials replaces conventional plastics in various applications, including packaging, consumer goods, and construction materials. Biodegradable, compostable, and bio-based alternatives derived from renewable resources are favored over fossil fuel-based plastics.
    - Innovations in material science, biotechnology, and green chemistry lead to the development of eco-

friendly materials with properties similar to conventional plastics but with reduced environmental impact and end-of-life considerations.

3. **Circular Economy Principles**:
   o A circular economy model is adopted where resources are managed sustainably, waste is minimized, and materials are reused, repaired, recycled, or repurposed at the end of their lifecycle.
   o Closed-loop systems for plastic recycling and recovery are established, enabling the collection, sorting, and processing of plastic waste into new products, packaging, or raw materials, thus reducing the demand for virgin plastics and minimizing environmental pollution.

4. **Innovations in Packaging and Design**:
   o Product redesign and eco-design principles prioritize packaging reduction, material efficiency, and recyclability. Packaging is designed to be minimal, lightweight, and easily recyclable or compostable, with a focus on extending product lifespan and minimizing environmental footprint.
   o Packaging innovations such as edible packaging, water-soluble films, and plant-based alternatives offer sustainable solutions to conventional plastic packaging and promote resource efficiency and waste reduction.

5. **Consumer Awareness and Behavior Change**:
   o Public awareness campaigns, education initiatives, and behavioral interventions raise awareness about the environmental impact of plastic pollution, promote sustainable consumption habits, and empower consumers to make informed choices.
   o Consumers prioritize products with minimal packaging, eco-friendly labeling, and sustainable certifications, and embrace reusable alternatives such as reusable bags, bottles, and containers in their daily lives.

6. **Policy and Regulatory Frameworks**:
   o Governments enact robust policies, regulations, and incentives to support the transition to a plastic-free future, including extended producer responsibility (EPR) schemes, plastic taxes, and product stewardship programs.

- International agreements and partnerships facilitate cross-border cooperation, harmonize standards, and promote best practices in plastic pollution prevention, waste management, and marine conservation.
7. **Corporate Responsibility and Industry Leadership:**
    - Businesses and industries take proactive steps to reduce their plastic footprint, adopt sustainable business practices, and invest in research and development of alternative materials and packaging solutions.
    - Corporate responsibility initiatives, voluntary commitments, and industry collaborations promote innovation, transparency, and accountability in plastic waste management and environmental stewardship.
8. **Community Engagement and Social Innovation:**
    - Local communities, grassroots organizations, and civil society groups play an active role in advocating for plastic-free policies, implementing community-led initiatives, and driving social innovation in waste reduction and resource management.
    - Community-based approaches such as zero-waste initiatives, plastic-free campaigns, and neighborhood clean-up events mobilize citizens, build social capital, and foster a sense of collective responsibility for environmental conservation.
9. **Global Collaboration and Solidarity:**
    - A plastic-free future requires global collaboration, solidarity, and shared responsibility among nations, stakeholders, and individuals to address the transboundary nature of plastic pollution and its impacts on ecosystems, wildlife, and human health.
    - International cooperation, knowledge exchange, and capacity-building efforts support developing countries in improving waste management infrastructure, implementing plastic pollution prevention measures, and promoting sustainable development pathways.
10. **Nature-Based Solutions and Ecological Restoration:**
    - Nature-based solutions such as wetland restoration, reforestation, and coastal ecosystem protection contribute to mitigating plastic pollution, enhancing

natural resilience, and promoting biodiversity conservation.
- Ecological restoration efforts focus on restoring degraded habitats, removing plastic waste from ecosystems, and restoring natural processes that regulate plastic accumulation and degradation in the environment.

Potential obstacles and solutions

Addressing plastic pollution and striving for a plastic-free future involves overcoming various obstacles that stem from social, economic, political, and technological factors. Identifying and addressing these obstacles is crucial for implementing effective solutions and achieving meaningful progress towards reducing plastic waste. Here's a detailed exploration of potential obstacles and solutions:

1. **Lack of Public Awareness and Engagement**:
   - Obstacle: Limited public awareness and engagement about the environmental impact of plastic pollution, as well as the importance of reducing plastic consumption and waste.
   - Solution: Implement comprehensive public awareness campaigns, education initiatives, and community outreach programs to raise awareness about the detrimental effects of plastic pollution on ecosystems, wildlife, and human health. Foster citizen engagement through participatory approaches, grassroots movements, and social media platforms to mobilize support for plastic-free initiatives.
2. **Inadequate Waste Management Infrastructure**:
   - Obstacle: Insufficient waste management infrastructure in many regions, particularly in developing countries, leading to inadequate collection, recycling, and disposal of plastic waste.
   - Solution: Invest in the development of robust waste management systems, including waste collection, sorting, recycling, and disposal facilities, to improve waste collection coverage and recycling rates. Provide technical assistance, capacity-building support, and financing mechanisms to enhance waste management infrastructure and build institutional capacity at the local, national, and regional levels.
3. **Plastic Industry Influence and Lobbying**:
   - Obstacle: Strong influence and lobbying efforts by the plastic industry to resist regulation, maintain market dominance, and promote continued plastic production and consumption.

- Solution: Strengthen regulatory frameworks, transparency measures, and accountability mechanisms to regulate the plastic industry, minimize conflicts of interest, and prevent undue influence on policymaking processes. Encourage industry collaboration, innovation, and corporate responsibility initiatives to promote sustainable practices, reduce plastic waste, and transition towards circular economy models.

4. **Complexity of Plastic Supply Chains**:
   - Obstacle: Complexity and fragmentation of plastic supply chains, which involve multiple stakeholders, producers, distributors, retailers, and consumers, making it challenging to track, trace, and regulate plastic production and distribution.
   - Solution: Implement supply chain transparency measures, traceability systems, and product labeling requirements to enhance visibility and accountability across the plastic value chain. Promote supply chain collaboration, information sharing, and best practices exchange to improve resource efficiency, reduce waste generation, and promote sustainable sourcing and production practices.

5. **Economic Dependence on Plastics**:
   - Obstacle: Economic dependence on plastics in various sectors, including packaging, manufacturing, construction, and healthcare, making it difficult to transition to alternative materials and sustainable practices.
   - Solution: Promote economic diversification, innovation, and investment in alternative materials, green technologies, and circular economy business models to reduce reliance on plastics and create new economic opportunities. Provide financial incentives, subsidies, and tax breaks for businesses that adopt sustainable practices, invest in eco-friendly alternatives, and prioritize waste reduction and recycling.

6. **Infrastructure and Technological Limitations**:
   - Obstacle: Technological limitations and infrastructure constraints, such as inadequate recycling technologies, limited access to recycling facilities, and lack of market demand for recycled

materials, hinder effective plastic waste management and recycling efforts.
- Solution: Invest in research and development of innovative recycling technologies, material recovery processes, and circular economy solutions to overcome technological barriers and improve the efficiency, scalability, and viability of plastic recycling. Expand and upgrade recycling infrastructure, collection systems, and end-markets for recycled materials to create a demand-driven recycling ecosystem that incentivizes investment and innovation in plastic waste management.

7. **Policy Fragmentation and Regulatory Gaps**:
   - Obstacle: Fragmentation and inconsistency of policies, regulations, and standards related to plastic pollution prevention, waste management, and recycling across different jurisdictions and sectors.
   - Solution: Harmonize and standardize regulatory frameworks, policy approaches, and best practices at the national, regional, and international levels to create a cohesive and coordinated response to plastic pollution. Strengthen enforcement mechanisms, compliance measures, and cross-border cooperation to address regulatory gaps, mitigate jurisdictional conflicts, and promote regulatory convergence on plastic waste management and pollution prevention.

8. **Social and Cultural Norms**:
   - Obstacle: Deeply ingrained social and cultural norms surrounding plastic use, convenience culture, and disposability mindset, which perpetuate unsustainable consumption patterns and hinder behavior change.
   - Solution: Foster a cultural shift towards sustainability, waste reduction, and environmental stewardship through education, media campaigns, and community engagement initiatives. Promote alternative narratives, values, and lifestyle choices that prioritize environmental conservation, resource efficiency, and conscious consumption. Encourage social innovation, peer-to-peer networks, and community-led initiatives that empower individuals and communities to adopt plastic-free practices and advocate for systemic change.

9. **Inequities and Social Justice Issues:**
    - Obstacle: Inequities and social justice issues related to plastic pollution, including disproportionate impacts on marginalized communities, low-income populations, and vulnerable groups who bear the brunt of plastic pollution and environmental degradation.
    - Solution: Address environmental justice concerns and equity considerations in plastic pollution mitigation strategies by ensuring equitable access to clean air, water, and environments for all communities. Incorporate principles of environmental justice, social equity, and community engagement into policy development, decision-making processes, and resource allocation to promote inclusive and equitable outcomes in plastic pollution prevention and management efforts.
10. **Global Cooperation and Multilateral Action:**
    - Obstacle: Limited global cooperation, coordination, and multilateral action on plastic pollution, with fragmented approaches and insufficient international commitments to address the transboundary nature of the issue.
    - Solution: Strengthen international cooperation, multilateral agreements, and global partnerships to address plastic pollution comprehensively and collaboratively. Enhance collaboration between governments, international organizations, civil society, academia, and the private sector to share knowledge, resources, and best practices, and mobilize collective action to tackle plastic pollution at the global scale. Promote cross-sectoral initiatives, knowledge exchange platforms, and funding mechanisms to support capacity-building, innovation, and technology transfer in plastic pollution prevention and management worldwide.

Call to action for individuals and organizations

A call to action for individuals and organizations is essential for mobilizing collective efforts and driving meaningful change in addressing plastic pollution and transitioning towards a plastic-free future. It urges individuals, communities, businesses, governments, and civil society organizations to take concrete steps, adopt sustainable practices, and advocate for systemic solutions to reduce plastic waste, protect the environment, and safeguard human health. Here's a detailed call to action for individuals and organizations:

1. **Reduce Single-Use Plastics:**
    - Individuals: Minimize the use of single-use plastics such as plastic bags, bottles, utensils, and straws in daily life. Opt for reusable alternatives, carry a reusable water bottle, coffee mug, and shopping bag, and refuse single-use plastics when possible.
    - Organizations: Implement policies and practices to reduce the use of single-use plastics in workplaces, events, and operations. Provide employees with reusable alternatives, promote waste reduction strategies, and encourage responsible consumption behaviors among staff and customers.
2. **Reuse and Recycle Plastic Waste:**
    - Individuals: Practice responsible waste management by segregating plastic waste, recycling materials whenever possible, and supporting community recycling programs. Reduce, reuse, and repurpose plastic items to extend their lifespan and minimize environmental impact.
    - Organizations: Establish recycling programs, waste separation systems, and circular economy initiatives to maximize plastic recovery, recycling, and reuse. Partner with recycling facilities, waste management companies, and industry stakeholders to improve collection, sorting, and processing of plastic waste.
3. **Support Sustainable Products and Businesses:**
    - Individuals: Choose products and brands that prioritize sustainability, eco-friendly packaging, and plastic-free alternatives. Support businesses and retailers that offer sustainable options, reduce

packaging waste, and promote ethical sourcing and production practices.
- Organizations: Source sustainable materials, packaging, and supplies for products and operations. Collaborate with suppliers, manufacturers, and distributors to transition to eco-friendly packaging, reduce plastic packaging waste, and adopt circular economy principles throughout the supply chain.

4. **Advocate for Policy Change**:
    - Individuals: Advocate for stronger policies, regulations, and legislation to address plastic pollution, promote waste reduction, and incentivize sustainable alternatives. Write to elected officials, support petitions, and participate in public consultations to voice concerns and demand action on plastic waste issues.
    - Organizations: Engage in policy advocacy efforts, industry coalitions, and stakeholder dialogues to influence government policies, shape regulatory frameworks, and support initiatives that promote plastic pollution prevention, waste management, and circular economy solutions.

5. **Educate and Raise Awareness**:
    - Individuals: Educate yourself and others about the environmental impact of plastic pollution, the importance of waste reduction, and the benefits of sustainable living practices. Share information, resources, and tips on social media, community forums, and educational platforms to raise awareness and inspire action.
    - Organizations: Conduct outreach and educational campaigns to raise awareness about plastic pollution, engage employees, customers, and stakeholders, and promote sustainable behavior change. Offer educational programs, workshops, and training sessions on plastic-free living, waste reduction, and environmental stewardship.

6. **Invest in Innovation and Research**:
    - Individuals: Support research, innovation, and entrepreneurship in developing alternative materials, recycling technologies, and sustainable solutions to plastic pollution. Invest in startups, crowdfunding campaigns, and research projects that focus on plastic

waste reduction, circular economy innovations, and eco-friendly alternatives.
- Organizations: Allocate resources, funding, and investments towards research and development of sustainable materials, packaging innovations, and waste management technologies. Partner with academic institutions, research organizations, and industry consortia to drive innovation, scale up solutions, and address key challenges in plastic pollution prevention and management.

7. **Engage in Community Action**:
   - Individuals: Participate in community clean-up events, volunteer programs, and grassroots initiatives that aim to remove plastic waste from natural environments, restore ecosystems, and raise awareness about local environmental issues.
   - Organizations: Support community-based conservation projects, environmental organizations, and NGO initiatives that focus on plastic pollution prevention, habitat restoration, and community engagement. Foster partnerships with local communities, indigenous groups, and civil society organizations to collaborate on shared environmental goals and initiatives.

8. **Promote Corporate Responsibility**:
   - Individuals: Hold corporations and businesses accountable for their plastic waste footprint, unsustainable practices, and environmental impacts. Support companies that demonstrate corporate responsibility, transparency, and commitment to sustainable development goals.
   - Organizations: Lead by example and adopt corporate responsibility initiatives, sustainability commitments, and environmental management practices that prioritize waste reduction, resource efficiency, and environmental stewardship. Implement eco-friendly policies, green procurement practices, and sustainability reporting mechanisms to track progress and accountability.

9. **Collaborate Across Sectors and Stakeholders**:
   - Individuals: Collaborate with diverse stakeholders, including government agencies, businesses, NGOs, academic institutions, and community groups, to

drive collective action and achieve shared goals in plastic pollution prevention and waste management.
- Organizations: Foster cross-sectoral collaborations, partnerships, and alliances that leverage complementary strengths, resources, and expertise to address complex challenges related to plastic pollution. Engage in multi-stakeholder initiatives, public-private partnerships, and industry collaborations to promote knowledge sharing, best practices exchange, and innovation in plastic waste reduction efforts.

10. **Lead by Example and Inspire Others**:
    - Individuals: Lead by example and inspire others to adopt plastic-free lifestyles, sustainable practices, and responsible consumption habits. Be a role model in your community, workplace, and social circles by demonstrating commitment to environmental stewardship and advocating for positive change.
    - Organizations: Demonstrate leadership and commitment to sustainability, environmental responsibility, and corporate citizenship by integrating plastic pollution prevention goals, targets, and metrics into organizational strategies, policies, and operations. Share success stories, case studies, and best practices to inspire and empower others to follow suit and join the movement towards a plastic-free future.

Urgency in ending plastic pollution

As we consider the urgency of ending plastic pollution and restoring ecosystems, it becomes clear that we are at a critical juncture in human history. The pervasive presence of plastic pollution poses significant threats to biodiversity, ecosystem health, and human well-being, necessitating immediate action to mitigate its adverse impacts and transition towards a more sustainable and resilient future. Here are some final thoughts on the urgency of this imperative:

1. **Environmental Crisis**: Plastic pollution has reached unprecedented levels, contaminating terrestrial and aquatic ecosystems, endangering wildlife, and disrupting natural processes. The accumulation of plastic waste in oceans, rivers, and landscapes threatens marine life, pollutes waterways, and compromises ecosystem services essential for human survival.
2. **Human Health Concerns**: Plastic pollution poses serious risks to human health, as toxic chemicals leach from plastic products, contaminate food and water sources, and enter the human food chain. Microplastics, nanoplastics, and plastic additives have been found in air, soil, water, and food, raising concerns about their potential health effects on human populations.
3. **Social and Economic Impacts**: Plastic pollution exacerbates social inequalities, disproportionately affecting marginalized communities, low-income populations, and vulnerable groups who bear the brunt of environmental degradation and pollution-related health risks. Plastic pollution also imposes significant economic costs, including cleanup expenses, tourism losses, and damage to fisheries and coastal economies.
4. **Ecosystem Degradation**: Plastic pollution threatens the integrity and resilience of ecosystems, degrading habitats, altering food webs, and disrupting ecological processes. Marine and terrestrial ecosystems suffer from entanglement, ingestion, and habitat degradation caused by plastic debris, leading to biodiversity loss, species extinction, and ecosystem collapse.
5. **Climate Change Connections**: The production, use, and disposal of plastics contribute to climate change through

greenhouse gas emissions, fossil fuel extraction, and energy-intensive manufacturing processes. Addressing plastic pollution and transitioning towards sustainable alternatives are essential components of climate change mitigation and adaptation strategies.
6. **Interconnected Challenges**: Plastic pollution is intrinsically linked to other environmental challenges, including climate change, biodiversity loss, deforestation, and habitat destruction. Addressing plastic pollution requires integrated solutions that address root causes, systemic drivers, and underlying social, economic, and political factors contributing to environmental degradation.
7. **Global Responsibility**: Plastic pollution is a global issue that transcends national borders, requiring collective action, shared responsibility, and international cooperation to address effectively. As stewards of the planet, we have a moral and ethical obligation to protect ecosystems, conserve biodiversity, and safeguard the health and well-being of present and future generations.
8. **Window of Opportunity**: While the challenges posed by plastic pollution are daunting, there is also a window of opportunity for transformative change. Advances in technology, innovation, policy, and public awareness provide pathways for solutions-oriented approaches that prioritize waste reduction, resource efficiency, and circular economy principles.
9. **Call to Action**: Ending plastic pollution and restoring ecosystems requires a concerted effort from individuals, communities, businesses, governments, and civil society organizations. It calls for bold leadership, collective responsibility, and sustained commitment to implementing evidence-based solutions, advocating for policy change, and driving systemic transformation across all sectors of society.
10. **Hope for the Future**: Despite the gravity of the challenges we face, there is hope for the future. By working together, harnessing our collective creativity, and drawing upon the resilience of nature, we can overcome the obstacles posed by plastic pollution and restore balance to our planet. Together, we can build a more sustainable, equitable, and plastic-free world for generations to come.

www.ingramcontent.com/pod-product-compliance
Lightning Source LLC
Chambersburg PA
CBHW070353230526
45471CB00006B/2540